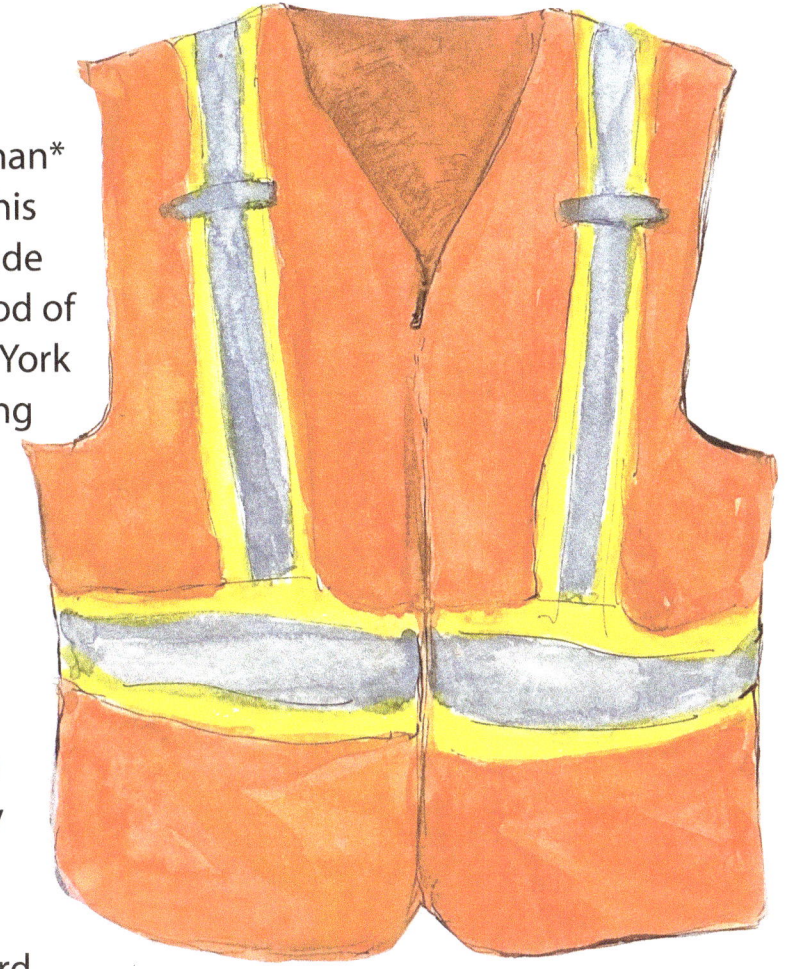

T he apprentices (seven women, and one man) and a journeywirewoman* who collaborated on this book are all members of the trade union, International Brotherhood of Electrical Workers Local 3, New York City. The apprentices are learning to become electricians in a 5 year rigorous training program which involves on-the-job training every day, electrical theory classes, and college courses where they earn an associates or bachelor's degree in Labor Studies from the Harry Van Arsdale Jr. School of Labor Studies, SUNY/Empire State College. The Joint Industry Board, made up of Local 3 and the National Electrical Contractors Association (NECA), through collective bargaining, covers the apprentices' college tuition, allowing them to graduate debt free.

This book originated in classes taught by Professor Sharon Szymanski, where the question emerged, "Why aren't there more women electricians?" A key answer was because young girls aren't aware that being a union electrician is a rewarding and lucrative career.

*Journeywirewoman is the designation on the union cards female union electricians receive upon completing their apprenticeship training. Union women electricians are very proud to be called Wirewomen. Many thanks to Journeywirewoman Erin Sullivan, who provided invaluable information and support for this book project.

We're all New York City WireWomen. We're superheroes and we want to tell you what we do and why we love our careers as electricians – **UNION** electricians.

We pull wire with the strength of an elephant tugging out tree trunks, we climb ladders with the agility of a mountain lion scaling a peak, and we read blueprints with the wisdom and inquisitiveness of an owl. We use our extraordinary powers to build and light up our city.

WireWomen's work is exciting and full of adventures. We even have helpers named **APPRENTICE** who get us coffee and doughnuts and hand us our tools and gadgets. In this book we'll share with you what we do and the tools we use.

WireWomen install the lights in baseball stadiums. We get to ride up in this huge machine called a knuckleboom and it's really fun. If electricians didn't light up the ballparks, the baseball players couldn't hit homeruns and the fans couldn't find their seats or put ketchup on their hotdogs.

"Hey APPRENTICE, 'Who's on First?'"

WireWomen work in the sky, where we help build skyscrapers. We have the best" office views." We're right up there with the gargoyles. We see falcons fly (really), and we can wave to people in airplanes. We can see forever. We can even see the tallest ferris wheels around the world.

Sometimes we even hang outside these buildings.

"Hey, APPRENTICE, we need a yo-yo here. No, not the toy." It's a harness with a slow release that lets us work safely from great heights. When we gear-up we feel like rock climbers.

We work inside skyscrapers, too. We provide the electricity to make sure revolving doors "spin" and turnstiles "turn" and fire alarms "alarm." Imagine what would happen if you were in a stalled revolving door with an impatient alligator. Or if Apprentice couldn't get through the turnstile with 20 cups of steaming hot coffee for 20 thirsty WireWomen? Or if the fire department didn't know there was a fire or where to go to save people.

WireWomen go to work and see buildings grow. We see more and more progress every day. As an electrician we see a job start from studs to walls, from temporary lights to running cables for fixtures. It's real and it's what makes a job come together. We're proud of working with our sisters to build these structures, and we love showing our families and friends the buildings we've worked on. Our kids are proud to see the buildings mommy helped build.

WireWomen keep the subways and trains running. We wire up the signals and track switchers so trains don't collide. We make sure all the emergency radios are working and that the temperature in the stations and trains is just right – hot in the summer and cold

"Hey APPRENTICE. We need a pony here. No, not the kind you ride."

We use ponies to put threads on a galvanized conduit. And just like a real pony, if you don't treat it right it will give you a kick.

in the winter ☺. We even run wires down the middle of the tracks to keep all the snow off. Sometimes we need to crawl into small spaces and pretend we're pretzels or a contortionist in the circus.

And Santa even borrows our tools – a hand bender – to put the "bend" in candy canes.

WireWomen help Santa find New York City. We string the lights on the Christmas tree in Rockefeller Center and make sure it twinkles so Santa can find his way. That's over 50,000 lights weighing 11 tons that would stretch out for 5 miles.

WireWomen make sure everyone has a Happy New Year. We wire up and install the New Year's Eve ball that drops at midnight in Times Square. If we don't get the timing just right, people can't blow whistles in stranger's faces, toss confetti, sing Auld Lang Syne, and kiss at midnight.

18

"Hurry, hurry **APPRENTICE,** we need 50,000 multiple LED lights, now!"

WireWomen keep the airports going. Planes can't take off or land if the runways aren't lit up. And

the Air Traffic Controllers rely on us for making sure their guidance systems are powered up. Just think what would happen at Baggage Claim if the conveyor belts stopped moving.

WireWomen love the Museum of Natural History as much as you do. The dinosaurs couldn't **ROAR** and

the gems wouldn't **GLOW** and the exhibits would be dark if we didn't provide the power.

WireWomen are highly skilled, problem solvers, and independent thinkers. We use our physical strength and knowledge of electricity and tools to "turn on the lights." We really have to focus while reading blueprints and wiring lighting fixtures. When the lights go on after we've wired them and there's no explosion we go, "phew."

Most of us never thought we'd be electricians. Even though some of us liked to use tools and build things or take apart computers when we were younger, we never thought we'd actually be an electrician – because that's not what we thought girls did. When we started, we didn't know how to use the simplest of tools – like a left-handed hammer. But after a year or so we're confident and proud of our skills. Often, time just flies by because the work is challenging and we're learning something new every day. Working with electricity is serious business so we need lots of training to keep everyone safe. Something that was so hard for us to pick up in the beginning is now something we love doing.

APPRENTICE says, "Give these WireWomen a trophy."

WireWomen are artists. We like to make our work look good because we take pride in what we do and are committed to doing a good job – it's a reflection of who we are as union-trained. What we think and what we do blend together – we choreograph bending pipes, splicing wires and landing fuses. We use our brains, eyes, ears, and touch to make sure everything is just right. And then…the magic happens.

All WireWomen carry electrical tape in their bags, in case of any emergency – on or off the job. "Hey pedestrian – let me tape up your drooping coat hem before you go to that job interview." "Geez, mummy, you're looking a little ragged. Let me wrap up that arm." "Yo, Bird, you got a boo boo on your beak. Got my tape right here."

APPRENTICE says, holding an open pizza box, "I've already got all the **BOY** tape we'll need for these emergencies – **Brown, Orange and Yellow.**"

WireWomen come from many different backgrounds and are of all diverse groups. It doesn't matter what type of family you belong to or whether you're younger or older – we are all sisters. We've been afterschool teachers and childcare workers; we've worked in offices, as graphic artists, in food service and as clerks in stores – all kinds of jobs that women typically hold. But we wanted to do something different and more exciting. And…that paid us a lot more.

APPRENTICE: "You know what? You look great in a hardhat and a high viz vest with a good union paycheck in your hand."

But we're **Union** WireWomen, which means we come together with other workers – our sisters and brothers – to get the best training, excellent health care and pensions, safe working conditions, and good pay (we can buy houses, and cars and save for vacations and our children's education). And, as union members, we do whatever we can to make the world a better place for everyone. WireWomen are superheroes and our union is our superpower.

Tools we use

Here's a list of some of the tools we use. We've also included some other ways these tools come in handy in our lives.

Deburring Tool

Used to smooth out sharp edges on conduit after cutting. This helps to protect the wires that will be pulled through the conduit. Can be used to file your nails or "depill" your sweaters.

Pump Pliers

Pliers with an adjustable setting to make the "jaw" of the pliers big or small to tighten or loosen different sized bolts. The grooves in the jaw provide a strong grip on any object. You can use it to reach your favorite book or toy on the top shelf in your closet.

Hammer Drill

Hammer drills are used to drill holes to mount materials on walls and ceilings and to drill through various surfaces like concrete floors. You can even use a hammer drill as a pogo stick (not advised).

Needlenose Pliers

Used to pull, twist, and shape wires in a tight spot like an outlet box. These pliers are perfect when your fingers are too big or bulky to get a good grip. Really useful to get small toys out of the sink drain.

Band Saw

An electric or battery powered saw with a long continuous sharp blade made up of sharp metal teeth stretched between two or more wheels to cut material. In a pinch, can quickly carve your holiday turkey, ham, fish, or tofu.

Spin Tight

This tool looks like a screwdriver but is used on hex head nuts. There is a different tool for each nut size. Can also be used to pit olives.

Snake

Also called "fish tape" because it's long and has a hook on the end. It's used to pull wire through conduit (pipe). You can use it to reel in dirty socks from under beds.

Level (Bubble Stick)

It makes sure the bends on conduit (metal tubing) and the boxes we mount are perfectly level. It has a magnet on the end that you can use to find coins in your couch.

Did You Know …?

Most electricians are men. In the United States only 2.4 percent of electricians are women. But their numbers are growing as they hear more about this career and all it has to offer.

Union electricians are well paid. A union electrician in New York City earns between $70,993 and $111,104 per year. Average yearly earnings for electricians in New York State are $81,340.

Women in all unionized jobs earn a whole lot more than women in non-union jobs. How much more? About $195 per week more. And, employers can't discriminate because the union makes sure all the workers are paid the same. Plus, union jobs have good health care, pension plans, and other benefits.

Source: https://www.nytimes.com/2020/03/05/business/women-electricians.html.

Source: https://www.comparably.com/salaries/salaries-for-union-electrician-in-new-york-ny. https://www.bls.gov/oes/current/oes_ny.htm#51-0000.

Source: Stronger Together, Union Membership Boosts Women's Earnings and Economic Security by Chuxuan Sun, Acadia Hall, and Elyse Shaw, Institute for Women's Policy Research. September 2021. https://iwpr.org/wp-content/uploads/2021/08/Stronger-Together-Union-Membership-Boosts-Womens-Earnings-and-Economic-Security_FINAL.pdf.

Apprentice authors

First Row (left to right) Britanny Rivera, Natalie Rivera, Daniel Emigholz, Samantha Granickas

Second Row (left to right) Sadhana Clarke, Joanna Kokosis, Theresa Murino, Mary Lin Gill

Third Row, Erin Sullivan, **journeywirewoman**

Lead author

SHARON SZYMANSKI is a professor of political economy and labor economics at the State University of New York (SUNY) Empire State College, the Harry Van Arsdale Jr. School of Labor Studies, New York City. She is a proud member of her union, United University Professions (UUP).

Illustrator

SETARE ARASHLOO is a multi-disciplinary artist who is interested in the intersections of art and social, political and environmental movements. She graduated with an MFA in Studio Art from Queens College and an Advanced Certificate in Critical Social Practice from Social Practice Queens (SPQ). Setare has worked as a museum and art educator in different institutions such as the Museum of Modern Art (MOMA), Queens College, and Empire State College (SUNY). Her collaborative and individual works have been exhibited internationally in the US, Iran, Afghanistan, France, Germany, and Australia. www.setarearashloo.com

Other Little Heroes Titles

The Cabbage That Came Back, Stephen Pearl (author)
Sara Pearl (translator), Rafael Pearl (illustrator)

Down on James Street, Nicole McCandless (author)
Byron Gramby (illustrator)

For All/Para Todos, Alejandra Domenzain (author)
Katherine Loh (illustrator) Irene Prieto de Coogan (translator)

Freedom Soldiers, (YA novel), Katherine Williams (author)

Good Guy Jake, Mark Torres (author)
Yana Muraskho (illustrator), Madelin Arroyo (translator)

Hats Off For Gabbie!, Marivir Montebon (author)
Yana Murashko (illustrator), Laura Flores (translator)

Jimmy's Carwash Adventure, Victor Narro (author)
Yana Murashko (illustrator), Madelin Arroyo (translator)

Joelito's Big Decision, Ann Berlak (author)
Daniel Camacho (illustrator), Jose Antonio Galloso (translator)

Manny & the Mango Tree, Ali Bustamante (author)
Monica Lunot-Kuker (illustrator), Mauricio Niebla (translator)

Margarito's Forest, Andy Carter (author)
Allison Havens (illustrator) Omar Majeia (translator)

Polar Bear Pete's Ice Is Melting! Timothy Sheard (author)
Kayla Fils-Aime, Byron Gramby (illustrators), Madelin Arroyo (translator)

Trailer Park, JC Dillard (author), Anna Usacheva (illustrator), Madelin Arroyo (translator)